The Chronnoisseur
Pocket Journal

The Chronnoisseur by Justin Klein

© 2018 by Justin Klein. All rights reserved.

No part of this book may be reproduced in any written, electronic, recording, or photocopying without written permission of the publisher or author. The exception would be in the case of brief quotations embodied in the critical articles or reviews and pages where permission is specifically granted by the publisher or author.

Although every precaution has been taken to verify the accuracy of the information contained herein, the author and publisher assume no responsibility for any errors or omissions. No liability is assumed for damages that may result from the use of information contained within.

www.lulu.com

Cover Design: Todd M. Schilling Design
Interior Design: Todd M. Schilling Design
Editor: Justin Klein

The purpose of this journal is to chronicle the consumption of cannabis for individuals who like to enjoy the different flavors, aromas, and methods of consumption available. I hope this book will encourage people to savor the variation between different strains as well as the variations between the same strains from different growers and regions. This journal will be a means for you to note the flavors, appearance, method of consumption, as well as the overall experience you like the best and least. Utilize this as a quick reference guide to determine whether you would like a future choice in product, and whether that product has unbeknownst to you already been consumed by you in the method of which you are currently about to try again; because, let's be honest, sometimes we forget.

Enjoy Responsibly!

CBD (Cannabidiol)/THC (Tetrahydrocannabinol) Basics

CBD (Cannabidiol): Known for non-psychoactive medical benefits. Does not bind to CB1 *(cannabinoid 1)* receptors found mostly in the brain.

Common uses: anti-tumoral, anti-inflammatory, anti-anxiety, anti-convulsant, painkiller, neuroprotectant.

THC *(Tetrahydrocannabinol):* Responsible for psychoactive effects of cannabis consumption. Binds with CB1 receptors.

Common uses: pain killer, muscle relaxant, sleep aid, relaxation, appetite stimulant

These statements have not been evaluated by the FDA and are not intended to diagnose, treat or cure any disease. Always check with your physician before starting a new dietary supplement program

* November 22, 2017. THC vs CBD: What's the Difference.
  Https://www.leafscience.com

Common Terpenes & Effects

Pinene (Alpha/Beta): Pine taste and aroma. Medical values include anti-inflammatory and bronchodilator properties. Common effects are alertness, memory retention, and anti-drowsiness.

Limonene: Citrus taste and aroma. Common effects include elevated mood and stress relief.

Linalool: Floral and sweet taste and aroma. Common effects include pain relief and sedative properties.

Myrcene: Musky/Herbal taste and aroma. Medical values include relaxing and sedative effects. Amount of myrcene present determines the indica and sativa effects of a strain.

Caryophyllene: Peppery/Spicy taste and aroma. Medical values include anti-inflammatory and pain relieving properties.

Humulene: Earthy/Hoppy taste and aroma. Medical values include appetite inducing and anti-inflammatory properties.

These statements have not been evaluated by the FDA and are not intended to diagnose, treat or cure any disease. Always check with your physician before starting a new dietary supplement program

*Rahm, Bailey (Feb 2014) Terpenes: The Flavors of Cannabis Aromatherapy. Https://www.leafly.com/news/cannabis-101

*Http://www.sclabs.com/terpenes

How to Fill Out

In the dispensary section list the name and/or location from which you purchased your cannabis. This will allow you to differentiate between strains in different regions and from dispensary/growers in the same region. The date section can be used as a reference to see whether the characteristic of your cannabis, or even your palate, has changed over time.

Next it is important to list the strain name of the cannabis you are consuming. Check whether the strain is an indica, sativa, or check both for hybrid. In the spaces provided next to the "THC" and "CBD" tabs; list the percentages, if available, or simply check which is dominant. The "Price" section is meant for you to note the difference in prices between regions, dispensaries, and strains.

The next few sections are meant help in describing the cannabis prior to consumption. Utilize the blank terpene wheel under the "Scent" tab to shade in the level of each terpene noted on the wheel. Circle what form of cannabis you will be consuming under the "Consumption Method" tab. The blank "Description" area of this section should be used to elaborate on everything that the terpene wheel is not fully expressing. For example, noting the consistency and/or clarity of concentrates and topicals.

Next is the section listing ailments that you may be using cannabis to remedy. Find the appropriate ailment, or ailments, that you are treating and mark the beginning point prior to consumption. If your particular ailment is not listed, use the bars labeled "Other" to log the change in your particular ailment.

These next sections will be used to describe the consumption of the cannabis. Fill out the terpene wheel under the "Taste" tab in the same manner as you did with the terpene wheel above. The blank "Description" area under this section should be utilized to elaborate on the type of method of consumption as well as the characteristics of the taste and smoke.

Lastly, capture your overall impression of the cannabis consumed. Revisit the ailments after consumption and log the end point to note how effective that particular strain or method of consumption was. Rate your overall experience on a scale of 1 through 10, and circle "Yes" or "No" depending on whether you would recommend this strain, and method, as a remedy to your ailments.

Under "Experience" you can describe a synopsis of what type of affect the cannabis had on you while under the influence of that particular strain, using that particular method.

The following pages are examples of how to complete the pages of your MCJ Log.

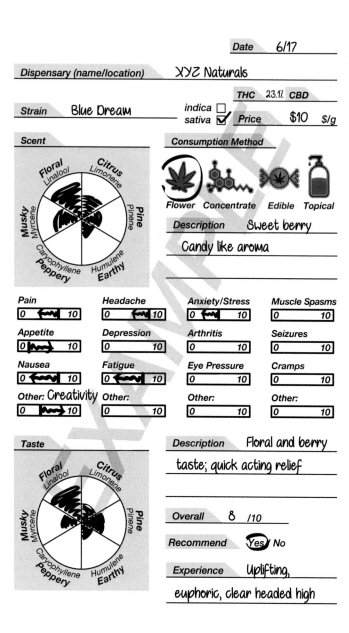

Date	10/17
Dispensary (name/location)	420 Alternative Medicine

THC ✓ CBD

Strain	MT Silvertip Live Resin	indica ✓ / sativa ☐	Price	$30 $/g

Scent

Floral (Linalool), Citrus (Limonene), Pine (Pinene), Earthy (Humulene), Peppery (Caryophyllene), Musky (Myrcene)

Consumption Method

Flower, **Concentrate** (circled), Edible, Topical

Description Sweet smell with piney and earthy tones grainy consistency

Pain	Headache	Anxiety/Stress	Muscle Spasms
0 ~~ 10	0 ~~ 10	0 10	0 10

Appetite	Depression	Arthritis	Seizures
0 ~~> 10	0 ~~ 10	0 ~~ 10	0 10

Nausea	Fatigue	Eye Pressure	Cramps
0 ~~ 10	0 10	0 10	0 10

Other:	Other:	Other:	Other:
0 10	0 10	0 10	0 10

Taste

Floral (Linalool), Citrus (Limonene), Pine (Pinene), Earthy (Humulene), Peppery (Caryophyllene), Musky (Myrcene)

Description Pine and earth tones more dominant when smoked

Overall 9 /10

Recommend Yes / No

Experience Extremely effective and fast acting

Date	8/17
Dispensary (name/location)	ABC Alternative Medicine
Strain	40mg Caramel
THC	40mg
CBD	
indica	✓
sativa	✓
Price	$5 $/g

Scent

Floral Linalool / *Citrus* Limonene / *Musky* Myrcene / *Pine* Pinene / *Peppery* Caryophyllene / *Earthy* Humulene

Consumption Method

Flower | Concentrate | **Edible** (circled) | Topical

Description Sweet smell
very faint cannabis scent

Pain 0 ~~~ 10 (mark near left)
Headache 0 ~~~ 10 (mark near left)
Anxiety/Stress 0 ~~~ 10 (mark near left)
Muscle Spasms 0 — 10

Appetite 0 ~~> 10 (mark toward right)
Depression 0 — 10
Arthritis 0 — 10
Seizures 0 — 10

Nausea 0 ~~ 10 (mark near left)
Fatigue 0 ~~ 10 (mark near middle)
Eye Pressure 0 ~~ 10 (mark near left)
Cramps 0 ~~ 10 (mark near left)

Other: 0 — 10
Other: 0 — 10
Other: 0 — 10
Other: 0 — 10

Taste

Floral Linalool / *Citrus* Limonene / *Musky* Myrcene / *Pine* Pinene / *Peppery* Caryophyllene / *Earthy* Humulene

Description very tasty with very little cannabis taste

Overall 8 /10

Recommend **Yes** / No

Experience 45min - 1hr to take effect; smooth body high

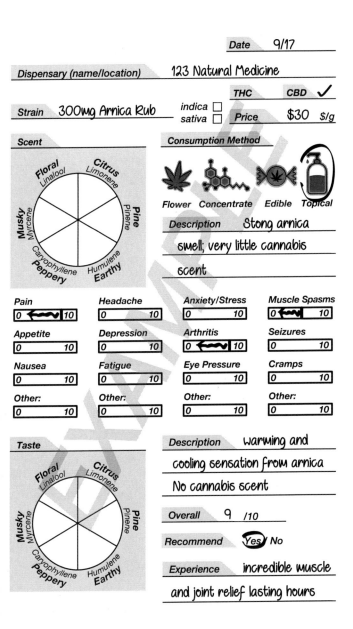

Date _____

Dispensary (name/location) _____

Strain _____   indica ☐   sativa ☐

THC _____ *CBD* _____

Price _____ $/g

Scent

Floral - Linalool
Citrus - Limonene
Pine - Pinene
Earthy - Humulene
Peppery - Caryophyllene
Musky - Myrcene

Consumption Method

Flower   Concentrate   Edible   Topical

Description _____
_____
_____

Pain	*Headache*	*Anxiety/Stress*	*Muscle Spasms*
0 — 10	0 — 10	0 — 10	0 — 10

Appetite	*Depression*	*Arthritis*	*Seizures*
0 — 10	0 — 10	0 — 10	0 — 10

Nausea	*Fatigue*	*Eye Pressure*	*Cramps*
0 — 10	0 — 10	0 — 10	0 — 10

Other:	*Other:*	*Other:*	*Other:*
0 — 10	0 — 10	0 — 10	0 — 10

Taste

Floral - Linalool
Citrus - Limonene
Pine - Pinene
Earthy - Humulene
Peppery - Caryophyllene
Musky - Myrcene

Description _____
_____
_____

Overall _____ /10

Recommend   Yes / No

Experience _____
_____

Date _____

Dispensary (name/location) _____

Strain _____

indica ☐
sativa ☐

THC ____ **CBD** ____

Price ____ $/g

Scent

- Floral — Linalool
- Citrus — Limonene
- Pine — Pinene
- Earthy — Humulene
- Peppery — Caryophyllene
- Musky — Myrcene

Consumption Method

Flower | Concentrate | Edible | Topical

Description _____
_____
_____

Pain [0 — 10]	**Headache** [0 — 10]	**Anxiety/Stress** [0 — 10]	**Muscle Spasms** [0 — 10]
Appetite [0 — 10]	**Depression** [0 — 10]	**Arthritis** [0 — 10]	**Seizures** [0 — 10]
Nausea [0 — 10]	**Fatigue** [0 — 10]	**Eye Pressure** [0 — 10]	**Cramps** [0 — 10]
Other: [0 — 10]	**Other:** [0 — 10]	**Other:** [0 — 10]	**Other:** [0 — 10]

Taste

- Floral — Linalool
- Citrus — Limonene
- Pine — Pinene
- Earthy — Humulene
- Peppery — Caryophyllene
- Musky — Myrcene

Description _____
_____

Overall ____ /10

Recommend Yes / No

Experience _____
_____

Date _____

Dispensary (name/location) _____

Strain _____   indica ☐   sativa ☐

THC _____ **CBD** _____ **Price** _____ $/g

Scent

Floral Linalool | *Citrus* Limonene
Musky Myrcene | *Pine* Pinene
Peppery Caryophyllene | *Earthy* Humulene

Consumption Method

Flower   Concentrate   Edible   Topical

Description _____

Pain	Headache	Anxiety/Stress	Muscle Spasms
0 — 10	0 — 10	0 — 10	0 — 10
Appetite	**Depression**	**Arthritis**	**Seizures**
0 — 10	0 — 10	0 — 10	0 — 10
Nausea	**Fatigue**	**Eye Pressure**	**Cramps**
0 — 10	0 — 10	0 — 10	0 — 10
Other:	**Other:**	**Other:**	**Other:**
0 — 10	0 — 10	0 — 10	0 — 10

Taste

Floral Linalool | *Citrus* Limonene
Musky Myrcene | *Pine* Pinene
Peppery Caryophyllene | *Earthy* Humulene

Description _____

Overall _____ /10

Recommend   Yes / No

Experience _____

Date _____

Dispensary (name/location) _____

Strain _____ indica ☐ sativa ☐

THC _____ **CBD** _____

Price _____ $/g

Scent

- Floral / Linalool
- Citrus / Limonene
- Pine / Pinene
- Earthy / Humulene
- Peppery / Caryophyllene
- Musky / Myrcene

Consumption Method

Flower   Concentrate   Edible   Topical

Description _____
_____
_____

Pain	Headache	Anxiety/Stress	Muscle Spasms
0 — 10	0 — 10	0 — 10	0 — 10

Appetite	Depression	Arthritis	Seizures
0 — 10	0 — 10	0 — 10	0 — 10

Nausea	Fatigue	Eye Pressure	Cramps
0 — 10	0 — 10	0 — 10	0 — 10

Other:	Other:	Other:	Other:
0 — 10	0 — 10	0 — 10	0 — 10

Taste

- Floral / Linalool
- Citrus / Limonene
- Pine / Pinene
- Earthy / Humulene
- Peppery / Caryophyllene
- Musky / Myrcene

Description _____
_____
_____

Overall _____ /10

Recommend Yes / No

Experience _____
_____

Date _____

Dispensary (name/location) _____

Strain _____ indica ☐ sativa ☐

THC _____ **CBD** _____
Price _____ $/g

Scent

Floral — Linalool
Citrus — Limonene
Musky — Myrcene
Pine — Pinene
Peppery — Caryophyllene
Earthy — Humulene

Consumption Method

Flower   Concentrate   Edible   Topical

Description _____
_____
_____

Pain	Headache	Anxiety/Stress	Muscle Spasms
0 — 10	0 — 10	0 — 10	0 — 10

Appetite	Depression	Arthritis	Seizures
0 — 10	0 — 10	0 — 10	0 — 10

Nausea	Fatigue	Eye Pressure	Cramps
0 — 10	0 — 10	0 — 10	0 — 10

Other:	Other:	Other:	Other:
0 — 10	0 — 10	0 — 10	0 — 10

Taste

Floral — Linalool
Citrus — Limonene
Musky — Myrcene
Pine — Pinene
Peppery — Caryophyllene
Earthy — Humulene

Description _____
_____
_____

Overall _____ /10

Recommend Yes / No

Experience _____
_____

Date _____

Dispensary (name/location) _____

| THC | CBD |

Strain _____ indica ☐ sativa ☐   Price _____ $/g

Scent

Floral — Linalool
Citrus — Limonene
Pine — Pinene
Musky — Myrcene
Caryophyllene — Peppery
Humulene — Earthy

Consumption Method

Flower   Concentrate   Edible   Topical

Description _____
_____
_____

Pain	Headache	Anxiety/Stress	Muscle Spasms
0 10	0 10	0 10	0 10
Appetite	Depression	Arthritis	Seizures
0 10	0 10	0 10	0 10
Nausea	Fatigue	Eye Pressure	Cramps
0 10	0 10	0 10	0 10
Other:	Other:	Other:	Other:
0 10	0 10	0 10	0 10

Taste

Floral — Linalool
Citrus — Limonene
Pine — Pinene
Musky — Myrcene
Caryophyllene — Peppery
Humulene — Earthy

Description _____
_____
_____

Overall _____ /10

Recommend   Yes / No

Experience _____
_____

Date _____

Dispensary (name/location) _____

Strain _____ indica ☐ sativa ☐

THC _____ CBD _____

Price _____ $/g

Scent

Floral — Linalool
Citrus — Limonene
Pine — Pinene
Humulene — Earthy
Caryophyllene — Peppery
Musky — Myrcene

Consumption Method

Flower   Concentrate   Edible   Topical

Description _____
_____
_____

Pain	*Headache*	*Anxiety/Stress*	*Muscle Spasms*
0 — 10	0 — 10	0 — 10	0 — 10

Appetite	*Depression*	*Arthritis*	*Seizures*
0 — 10	0 — 10	0 — 10	0 — 10

Nausea	*Fatigue*	*Eye Pressure*	*Cramps*
0 — 10	0 — 10	0 — 10	0 — 10

Other:	*Other:*	*Other:*	*Other:*
0 — 10	0 — 10	0 — 10	0 — 10

Taste

Floral — Linalool
Citrus — Limonene
Pine — Pinene
Humulene — Earthy
Caryophyllene — Peppery
Musky — Myrcene

Description _____
_____
_____

Overall _____ /10

Recommend   Yes / No

Experience _____
_____

Date _____

Dispensary (name/location) _____

Strain _____ indica ☐ sativa ☐

THC _____ CBD _____

Price _____ $/g

Scent

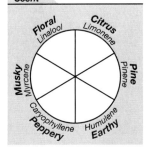

Consumption Method

Flower   Concentrate   Edible   Topical

Description _____
_____
_____

Pain
| 0 | 10 |

Headache
| 0 | 10 |

Anxiety/Stress
| 0 | 10 |

Muscle Spasms
| 0 | 10 |

Appetite
| 0 | 10 |

Depression
| 0 | 10 |

Arthritis
| 0 | 10 |

Seizures
| 0 | 10 |

Nausea
| 0 | 10 |

Fatigue
| 0 | 10 |

Eye Pressure
| 0 | 10 |

Cramps
| 0 | 10 |

Other:
| 0 | 10 |

Other:
| 0 | 10 |

Other:
| 0 | 10 |

Other:
| 0 | 10 |

Taste

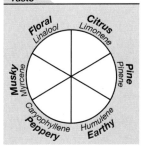

Description _____
_____

Overall _____ /10

Recommend   Yes / No

Experience _____
_____

Date _____

Dispensary (name/location) _____

Strain _____ indica ☐ sativa ☐

THC _____ **CBD** _____

Price _____ $/g

Scent

- Floral / Linalool
- Citrus / Limonene
- Pine / Pinene
- Earthy / Humulene
- Peppery / Caryophyllene
- Musky / Myrcene

Consumption Method

Flower Concentrate Edible Topical

Description _____

Pain	Headache	Anxiety/Stress	Muscle Spasms
0 — 10	0 — 10	0 — 10	0 — 10

Appetite	Depression	Arthritis	Seizures
0 — 10	0 — 10	0 — 10	0 — 10

Nausea	Fatigue	Eye Pressure	Cramps
0 — 10	0 — 10	0 — 10	0 — 10

Other:	Other:	Other:	Other:
0 — 10	0 — 10	0 — 10	0 — 10

Taste

- Floral / Linalool
- Citrus / Limonene
- Pine / Pinene
- Earthy / Humulene
- Peppery / Caryophyllene
- Musky / Myrcene

Description _____

Overall _____ /10

Recommend Yes / No

Experience _____

Date _____

Dispensary (name/location) _____

Strain _____ indica ☐ sativa ☐

THC ____ CBD ____

Price ____ $/g

Scent

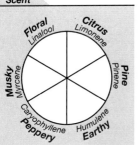

Consumption Method

Flower Concentrate Edible Topical

Description _____
_____
_____

Pain	Headache	Anxiety/Stress	Muscle Spasms
0 — 10	0 — 10	0 — 10	0 — 10

Appetite	Depression	Arthritis	Seizures
0 — 10	0 — 10	0 — 10	0 — 10

Nausea	Fatigue	Eye Pressure	Cramps
0 — 10	0 — 10	0 — 10	0 — 10

Other:	Other:	Other:	Other:
0 — 10	0 — 10	0 — 10	0 — 10

Taste

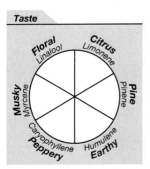

Description _____
_____
_____

Overall ____ /10

Recommend Yes / No

Experience _____
_____

Date _____

Dispensary (name/location) _____

Strain _____

indica ☐
sativa ☐

THC _____ CBD _____

Price _____ $/g

Scent

Floral — Linalool
Citrus — Limonene
Pine — Pinene
Earthy — Humulene
Peppery — Caryophyllene
Musky — Myrcene

Consumption Method

Flower   Concentrate   Edible   Topical

Description _____
_____
_____

Pain	Headache	Anxiety/Stress	Muscle Spasms
0 — 10	0 — 10	0 — 10	0 — 10

Appetite	Depression	Arthritis	Seizures
0 — 10	0 — 10	0 — 10	0 — 10

Nausea	Fatigue	Eye Pressure	Cramps
0 — 10	0 — 10	0 — 10	0 — 10

Other:	Other:	Other:	Other:
0 — 10	0 — 10	0 — 10	0 — 10

Taste

Floral — Linalool
Citrus — Limonene
Pine — Pinene
Earthy — Humulene
Peppery — Caryophyllene
Musky — Myrcene

Description _____
_____
_____

Overall _____ /10

Recommend   Yes / No

Experience _____
_____

Date _____

Dispensary (name/location) _____

THC	CBD

Strain _____ indica ☐ sativa ☐

Price _____ $/g

Scent

- Floral — Linalool
- Citrus — Limonene
- Pine — Pinene
- Earthy — Humulene
- Peppery — Caryophyllene
- Musky — Myrcene

Consumption Method

Flower   Concentrate   Edible   Topical

Description _____
_____
_____

Pain	Headache	Anxiety/Stress	Muscle Spasms
0 — 10	0 — 10	0 — 10	0 — 10

Appetite	Depression	Arthritis	Seizures
0 — 10	0 — 10	0 — 10	0 — 10

Nausea	Fatigue	Eye Pressure	Cramps
0 — 10	0 — 10	0 — 10	0 — 10

Other:	Other:	Other:	Other:
0 — 10	0 — 10	0 — 10	0 — 10

Taste

- Floral — Linalool
- Citrus — Limonene
- Pine — Pinene
- Earthy — Humulene
- Peppery — Caryophyllene
- Musky — Myrcene

Description _____
_____
_____

Overall ___ /10

Recommend   Yes / No

Experience _____
_____

Date _____

Dispensary (name/location) _____

Strain _____ indica ☐ sativa ☐

THC _____ *CBD* _____

Price _____ $/g

Scent

- Floral — Linalool
- Citrus — Limonene
- Pine — Pinene
- Earthy — Humulene
- Peppery — Caryophyllene
- Musky — Myrcene

Consumption Method

Flower Concentrate Edible Topical

Description
_____
_____
_____

Pain	Headache	Anxiety/Stress	Muscle Spasms
0 — 10	0 — 10	0 — 10	0 — 10
Appetite	**Depression**	**Arthritis**	**Seizures**
0 — 10	0 — 10	0 — 10	0 — 10
Nausea	**Fatigue**	**Eye Pressure**	**Cramps**
0 — 10	0 — 10	0 — 10	0 — 10
Other:	**Other:**	**Other:**	**Other:**
0 — 10	0 — 10	0 — 10	0 — 10

Taste

- Floral — Linalool
- Citrus — Limonene
- Pine — Pinene
- Earthy — Humulene
- Peppery — Caryophyllene
- Musky — Myrcene

Description
_____
_____

Overall _____ /10

Recommend Yes / No

Experience
_____
_____

Date _____

Dispensary (name/location) _____

Strain _____ indica ☐ sativa ☐

THC _____ **CBD** _____

Price _____ $/g

Scent

- Floral / Linalool
- Citrus / Limonene
- Pine / Pinene
- Earthy / Humulene
- Peppery / Caryophyllene
- Musky / Myrcene

Consumption Method

Flower   Concentrate   Edible   Topical

Description _____

Symptom	0–10
Pain	0 – 10
Headache	0 – 10
Anxiety/Stress	0 – 10
Muscle Spasms	0 – 10
Appetite	0 – 10
Depression	0 – 10
Arthritis	0 – 10
Seizures	0 – 10
Nausea	0 – 10
Fatigue	0 – 10
Eye Pressure	0 – 10
Cramps	0 – 10
Other:	0 – 10
Other:	0 – 10
Other:	0 – 10
Other:	0 – 10

Taste

- Floral / Linalool
- Citrus / Limonene
- Pine / Pinene
- Earthy / Humulene
- Peppery / Caryophyllene
- Musky / Myrcene

Description _____

Overall ____ /10

Recommend Yes / No

Experience _____

Date _____

Dispensary (name/location) _____

Strain _____ indica ☐ sativa ☐

THC _____ **CBD** _____
Price _____ $/g

Scent

Floral — Linalool
Citrus — Limonene
Pine — Pinene
Earthy — Humulene
Peppery — Caryophyllene
Musky — Myrcene

Consumption Method

Flower   Concentrate   Edible   Topical

Description _____
_____
_____

Pain	Headache	Anxiety/Stress	Muscle Spasms
0 — 10	0 — 10	0 — 10	0 — 10

Appetite	Depression	Arthritis	Seizures
0 — 10	0 — 10	0 — 10	0 — 10

Nausea	Fatigue	Eye Pressure	Cramps
0 — 10	0 — 10	0 — 10	0 — 10

Other:	Other:	Other:	Other:
0 — 10	0 — 10	0 — 10	0 — 10

Taste

Floral — Linalool
Citrus — Limonene
Pine — Pinene
Earthy — Humulene
Peppery — Caryophyllene
Musky — Myrcene

Description _____
_____
_____

Overall _____ /10

Recommend Yes / No

Experience _____
_____

Date _____

Dispensary (name/location) _____

Strain _____ indica ☐ sativa ☐

THC _____ **CBD** _____ **Price** _____ $/g

Scent

- Floral — Linalool
- Citrus — Limonene
- Pine — Pinene
- Earthy — Humulene
- Peppery — Caryophyllene
- Musky — Myrcene

Consumption Method

Flower   Concentrate   Edible   Topical

Description _____

Pain	Headache	Anxiety/Stress	Muscle Spasms
0 — 10	0 — 10	0 — 10	0 — 10
Appetite	**Depression**	**Arthritis**	**Seizures**
0 — 10	0 — 10	0 — 10	0 — 10
Nausea	**Fatigue**	**Eye Pressure**	**Cramps**
0 — 10	0 — 10	0 — 10	0 — 10
Other:	**Other:**	**Other:**	**Other:**
0 — 10	0 — 10	0 — 10	0 — 10

Taste

- Floral — Linalool
- Citrus — Limonene
- Pine — Pinene
- Earthy — Humulene
- Peppery — Caryophyllene
- Musky — Myrcene

Description _____

Overall ____ /10

Recommend Yes / No

Experience _____

Date _____

Dispensary (name/location) _____

Strain _____ indica ☐ sativa ☐

THC _____ **CBD** _____

Price _____ $/g

Scent

- Floral / Linalool
- Citrus / Limonene
- Pine / Pinene
- Earthy / Humulene
- Peppery / Caryophyllene
- Musky / Myrcene

Consumption Method

Flower Concentrate Edible Topical

Description _____

Pain	Headache	Anxiety/Stress	Muscle Spasms
0 – 10	0 – 10	0 – 10	0 – 10

Appetite	Depression	Arthritis	Seizures
0 – 10	0 – 10	0 – 10	0 – 10

Nausea	Fatigue	Eye Pressure	Cramps
0 – 10	0 – 10	0 – 10	0 – 10

Other:	Other:	Other:	Other:
0 – 10	0 – 10	0 – 10	0 – 10

Taste

- Floral / Linalool
- Citrus / Limonene
- Pine / Pinene
- Earthy / Humulene
- Peppery / Caryophyllene
- Musky / Myrcene

Description _____

Overall _____ /10

Recommend Yes / No

Experience _____

Date _____

Dispensary (name/location) _____

Strain _____ indica ☐ sativa ☐

THC _____ **CBD** _____ **Price** _____ $/g

Scent

- Floral / Linalool
- Citrus / Limonene
- Pine / Pinene
- Earthy / Humulene
- Peppery / Caryophyllene
- Musky / Myrcene

Consumption Method

Flower   Concentrate   Edible   Topical

Description _____

Pain	Headache	Anxiety/Stress	Muscle Spasms
0 — 10	0 — 10	0 — 10	0 — 10

Appetite	Depression	Arthritis	Seizures
0 — 10	0 — 10	0 — 10	0 — 10

Nausea	Fatigue	Eye Pressure	Cramps
0 — 10	0 — 10	0 — 10	0 — 10

Other:	Other:	Other:	Other:
0 — 10	0 — 10	0 — 10	0 — 10

Taste

- Floral / Linalool
- Citrus / Limonene
- Pine / Pinene
- Earthy / Humulene
- Peppery / Caryophyllene
- Musky / Myrcene

Description _____

Overall _____ /10

Recommend Yes / No

Experience _____

Date _____

Dispensary (name/location) _____

Strain _____ indica ☐ sativa ☐

THC _____ **CBD** _____
Price _____ $/g

Scent

Floral - Linalool
Citrus - Limonene
Pine - Pinene
Earthy - Humulene
Peppery - Caryophyllene
Musky - Myrcene

Consumption Method

Flower   Concentrate   Edible   Topical

Description
_____
_____
_____

Pain	Headache	Anxiety/Stress	Muscle Spasms
0 — 10	0 — 10	0 — 10	0 — 10

Appetite	Depression	Arthritis	Seizures
0 — 10	0 — 10	0 — 10	0 — 10

Nausea	Fatigue	Eye Pressure	Cramps
0 — 10	0 — 10	0 — 10	0 — 10

Other:	Other:	Other:	Other:
0 — 10	0 — 10	0 — 10	0 — 10

Taste

Floral - Linalool
Citrus - Limonene
Pine - Pinene
Earthy - Humulene
Peppery - Caryophyllene
Musky - Myrcene

Description
_____
_____

Overall _____ /10

Recommend   Yes / No

Experience
_____
_____

Date _____

Dispensary (name/location) _____

Strain _____ indica ☐ sativa ☐

THC _____ **CBD** _____

Price _____ $/g

Scent

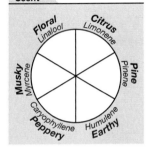

Consumption Method

Flower   Concentrate   Edible   Topical

Description _____
_____
_____

Pain	Headache	Anxiety/Stress	Muscle Spasms
0 10	0 10	0 10	0 10

Appetite	Depression	Arthritis	Seizures
0 10	0 10	0 10	0 10

Nausea	Fatigue	Eye Pressure	Cramps
0 10	0 10	0 10	0 10

Other:	Other:	Other:	Other:
0 10	0 10	0 10	0 10

Taste

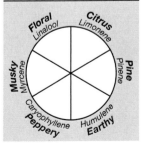

Description _____
_____
_____

Overall _____ /10

Recommend Yes / No

Experience _____
_____

Date _____

Dispensary (name/location) _____

Strain _____ indica ☐ sativa ☐

THC	CBD

Price _____ $/g

Scent

Floral — Linalool
Citrus — Limonene
Pine — Pinene
Earthy — Humulene
Peppery — Caryophyllene
Musky — Myrcene

Consumption Method

Flower Concentrate Edible Topical

Description
_____
_____
_____

Pain	Headache	Anxiety/Stress	Muscle Spasms
0 — 10	0 — 10	0 — 10	0 — 10
Appetite	**Depression**	**Arthritis**	**Seizures**
0 — 10	0 — 10	0 — 10	0 — 10
Nausea	**Fatigue**	**Eye Pressure**	**Cramps**
0 — 10	0 — 10	0 — 10	0 — 10
Other:	**Other:**	**Other:**	**Other:**
0 — 10	0 — 10	0 — 10	0 — 10

Taste

Floral — Linalool
Citrus — Limonene
Pine — Pinene
Earthy — Humulene
Peppery — Caryophyllene
Musky — Myrcene

Description
_____
_____
_____

Overall _____ /10

Recommend   Yes / No

Experience
_____
_____

Date _____

Dispensary (name/location) _____

THC	CBD

Strain _____ indica ☐ sativa ☐ **Price** _____ $/g

Scent

- Floral — Linalool
- Citrus — Limonene
- Pine — Pinene
- Earthy — Humulene
- Peppery — Caryophyllene
- Musky — Myrcene

Consumption Method

Flower Concentrate Edible Topical

Description _____
_____
_____

Pain	Headache	Anxiety/Stress	Muscle Spasms
0 — 10	0 — 10	0 — 10	0 — 10

Appetite	Depression	Arthritis	Seizures
0 — 10	0 — 10	0 — 10	0 — 10

Nausea	Fatigue	Eye Pressure	Cramps
0 — 10	0 — 10	0 — 10	0 — 10

Other:	Other:	Other:	Other:
0 — 10	0 — 10	0 — 10	0 — 10

Taste

- Floral — Linalool
- Citrus — Limonene
- Pine — Pinene
- Earthy — Humulene
- Peppery — Caryophyllene
- Musky — Myrcene

Description _____
_____
_____

Overall _____ /10

Recommend Yes / No

Experience _____
_____

Date _____

Dispensary (name/location) _____

Strain _____ indica ☐ sativa ☐

THC _____ **CBD** _____

Price _____ $/g

Scent

Floral Linalool
Citrus Limonene
Pine Pinene
Earthy Humulene
Peppery Caryophyllene
Musky Myrcene

Consumption Method

Flower   Concentrate   Edible   Topical

Description _____
_____
_____

Pain	Headache	Anxiety/Stress	Muscle Spasms
0 — 10	0 — 10	0 — 10	0 — 10

Appetite	Depression	Arthritis	Seizures
0 — 10	0 — 10	0 — 10	0 — 10

Nausea	Fatigue	Eye Pressure	Cramps
0 — 10	0 — 10	0 — 10	0 — 10

Other:	Other:	Other:	Other:
0 — 10	0 — 10	0 — 10	0 — 10

Taste

Floral Linalool
Citrus Limonene
Pine Pinene
Earthy Humulene
Peppery Caryophyllene
Musky Myrcene

Description _____
_____
_____

Overall _____ /10

Recommend Yes / No

Experience _____
_____

Date _____

Dispensary (name/location) _____

	THC	CBD
Strain _____ indica ☐ sativa ☐		
	Price	$/g

Scent

Floral — Linalool
Citrus — Limonene
Pine — Pinene
Earthy — Humulene
Peppery — Caryophyllene
Musky — Myrcene

Consumption Method

Flower   Concentrate   Edible   Topical

Description _____
_____
_____

Pain	Headache	Anxiety/Stress	Muscle Spasms
0 — 10	0 — 10	0 — 10	0 — 10

Appetite	Depression	Arthritis	Seizures
0 — 10	0 — 10	0 — 10	0 — 10

Nausea	Fatigue	Eye Pressure	Cramps
0 — 10	0 — 10	0 — 10	0 — 10

Other:	Other:	Other:	Other:
0 — 10	0 — 10	0 — 10	0 — 10

Taste

Floral — Linalool
Citrus — Limonene
Pine — Pinene
Earthy — Humulene
Peppery — Caryophyllene
Musky — Myrcene

Description _____
_____
_____

Overall _____ /10

Recommend Yes / No

Experience _____
_____

Date

Dispensary (name/location)

Strain indica ☐ sativa ☐

THC **CBD**

Price $/g

Scent

- Floral — Linalool
- Citrus — Limonene
- Pine — Pinene
- Earthy — Humulene
- Peppery — Caryophyllene
- Musky — Myrcene

Consumption Method

Flower Concentrate Edible Topical

Description

Pain	Headache	Anxiety/Stress	Muscle Spasms
0 — 10	0 — 10	0 — 10	0 — 10
Appetite	**Depression**	**Arthritis**	**Seizures**
0 — 10	0 — 10	0 — 10	0 — 10
Nausea	**Fatigue**	**Eye Pressure**	**Cramps**
0 — 10	0 — 10	0 — 10	0 — 10
Other:	**Other:**	**Other:**	**Other:**
0 — 10	0 — 10	0 — 10	0 — 10

Taste

- Floral — Linalool
- Citrus — Limonene
- Pine — Pinene
- Earthy — Humulene
- Peppery — Caryophyllene
- Musky — Myrcene

Description

Overall /10

Recommend Yes / No

Experience

Date _____

Dispensary (name/location) _____

Strain _____   indica ☐   **THC** ____ **CBD** ____
                              sativa ☐   **Price** ____ $/g

Scent

Floral — Linalool
Citrus — Limonene
Pine — Pinene
Musky — Myrcene
Peppery — Caryophyllene
Earthy — Humulene

Consumption Method

Flower   Concentrate   Edible   Topical

Description _____
_____
_____

Pain	Headache	Anxiety/Stress	Muscle Spasms
0 — 10	0 — 10	0 — 10	0 — 10

Appetite	Depression	Arthritis	Seizures
0 — 10	0 — 10	0 — 10	0 — 10

Nausea	Fatigue	Eye Pressure	Cramps
0 — 10	0 — 10	0 — 10	0 — 10

Other:	Other:	Other:	Other:
0 — 10	0 — 10	0 — 10	0 — 10

Taste

Floral — Linalool
Citrus — Limonene
Pine — Pinene
Musky — Myrcene
Peppery — Caryophyllene
Earthy — Humulene

Description _____
_____

Overall ____ /10

Recommend   Yes / No

Experience _____
_____

Date _____

Dispensary (name/location) _____

Strain _____

indica ☐
sativa ☐

THC	CBD

Price _____ $/g

Scent

- Floral — Linalool
- Citrus — Limonene
- Pine — Pinene
- Earthy — Humulene
- Peppery — Caryophyllene
- Musky — Myrcene

Consumption Method

Flower   **Concentrate**   **Edible**   **Topical**

Description _____
_____
_____

Pain	Headache	Anxiety/Stress	Muscle Spasms
0 — 10	0 — 10	0 — 10	0 — 10

Appetite	Depression	Arthritis	Seizures
0 — 10	0 — 10	0 — 10	0 — 10

Nausea	Fatigue	Eye Pressure	Cramps
0 — 10	0 — 10	0 — 10	0 — 10

Other:	Other:	Other:	Other:
0 — 10	0 — 10	0 — 10	0 — 10

Taste

- Floral — Linalool
- Citrus — Limonene
- Pine — Pinene
- Earthy — Humulene
- Peppery — Caryophyllene
- Musky — Myrcene

Description _____
_____

Overall _____ /10

Recommend   Yes / No

Experience _____
_____

Date _____

Dispensary (name/location) _____

Strain _____ indica ☐ sativa ☐

THC _____ **CBD** _____

Price _____ $/g

Scent

- Floral / Linalool
- Citrus / Limonene
- Pine / Pinene
- Earthy / Humulene
- Peppery / Caryophyllene
- Musky / Myrcene

Consumption Method

Flower Concentrate Edible Topical

Description _____

Pain	Headache	Anxiety/Stress	Muscle Spasms
0 — 10	0 — 10	0 — 10	0 — 10
Appetite	**Depression**	**Arthritis**	**Seizures**
0 — 10	0 — 10	0 — 10	0 — 10
Nausea	**Fatigue**	**Eye Pressure**	**Cramps**
0 — 10	0 — 10	0 — 10	0 — 10
Other:	**Other:**	**Other:**	**Other:**
0 — 10	0 — 10	0 — 10	0 — 10

Taste

- Floral / Linalool
- Citrus / Limonene
- Pine / Pinene
- Earthy / Humulene
- Peppery / Caryophyllene
- Musky / Myrcene

Description _____

Overall _____ /10

Recommend Yes / No

Experience _____

Date _____

Dispensary (name/location) _____

Strain _____ indica ☐ sativa ☐

THC _____ **CBD** _____

Price _____ $/g

Scent

- Floral / Linalool
- Citrus / Limonene
- Pine / Pinene
- Earthy / Humulene
- Peppery / Caryophyllene
- Musky / Myrcene

Consumption Method

Flower Concentrate Edible Topical

Description _____
_____
_____

Pain	Headache	Anxiety/Stress	Muscle Spasms
0 — 10	0 — 10	0 — 10	0 — 10

Appetite	Depression	Arthritis	Seizures
0 — 10	0 — 10	0 — 10	0 — 10

Nausea	Fatigue	Eye Pressure	Cramps
0 — 10	0 — 10	0 — 10	0 — 10

Other:	Other:	Other:	Other:
0 — 10	0 — 10	0 — 10	0 — 10

Taste

- Floral / Linalool
- Citrus / Limonene
- Pine / Pinene
- Earthy / Humulene
- Peppery / Caryophyllene
- Musky / Myrcene

Description _____
_____
_____

Overall _____ /10

Recommend Yes / No

Experience _____
_____

Date _____

Dispensary (name/location) _____

Strain _____ indica ☐ sativa ☐

THC ____ CBD ____

Price ____ $/g

Scent

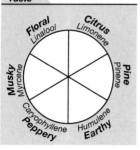

Consumption Method

Flower **Concentrate** **Edible** **Topical**

Description _____
_____
_____

Pain	Headache	Anxiety/Stress	Muscle Spasms
0 — 10	0 — 10	0 — 10	0 — 10

Appetite	Depression	Arthritis	Seizures
0 — 10	0 — 10	0 — 10	0 — 10

Nausea	Fatigue	Eye Pressure	Cramps
0 — 10	0 — 10	0 — 10	0 — 10

Other:	Other:	Other:	Other:
0 — 10	0 — 10	0 — 10	0 — 10

Taste

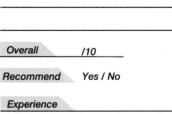

Description _____
_____
_____

Overall ____ /10

Recommend Yes / No

Experience _____
_____

Date _____

Dispensary (name/location) _____

Strain _____ indica ☐ sativa ☐

THC ____ CBD ____

Price ____ $/g

Scent

- Floral — Linalool
- Citrus — Limonene
- Musky — Myrcene
- Pine — Pinene
- Peppery — Caryophyllene
- Earthy — Humulene

Consumption Method

Flower   Concentrate   Edible   Topical

Description _____

Pain	Headache	Anxiety/Stress	Muscle Spasms
0 — 10	0 — 10	0 — 10	0 — 10
Appetite	**Depression**	**Arthritis**	**Seizures**
0 — 10	0 — 10	0 — 10	0 — 10
Nausea	**Fatigue**	**Eye Pressure**	**Cramps**
0 — 10	0 — 10	0 — 10	0 — 10
Other:	**Other:**	**Other:**	**Other:**
0 — 10	0 — 10	0 — 10	0 — 10

Taste

- Floral — Linalool
- Citrus — Limonene
- Musky — Myrcene
- Pine — Pinene
- Peppery — Caryophyllene
- Earthy — Humulene

Description _____

Overall ____ /10

Recommend Yes / No

Experience _____

Date _____

Dispensary (name/location) _____

Strain _____ indica ☐ sativa ☐

THC ____ *CBD* ____

Price ____ $/g

Scent

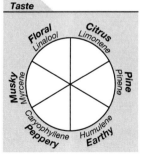

Consumption Method

🌿 Flower   ⚛ Concentrate   🍬 Edible   🧴 Topical

Description _____
_____
_____

Pain	Headache	Anxiety/Stress	Muscle Spasms
0 — 10	0 — 10	0 — 10	0 — 10

Appetite	Depression	Arthritis	Seizures
0 — 10	0 — 10	0 — 10	0 — 10

Nausea	Fatigue	Eye Pressure	Cramps
0 — 10	0 — 10	0 — 10	0 — 10

Other:	Other:	Other:	Other:
0 — 10	0 — 10	0 — 10	0 — 10

Taste

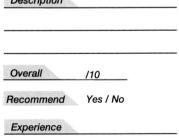

Description _____
_____
_____

Overall ____ /10

Recommend Yes / No

Experience _____
_____

Date _____

Dispensary (name/location) _____

Strain _____ indica ☐ sativa ☐

THC ___ *CBD* ___

Price ___ $/g

Scent

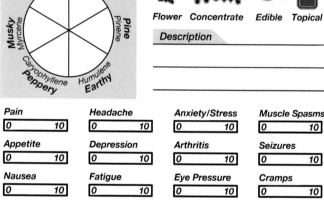

Consumption Method

Flower Concentrate Edible Topical

Description _____
_____
_____

Pain	Headache	Anxiety/Stress	Muscle Spasms
0 — 10	0 — 10	0 — 10	0 — 10

Appetite	Depression	Arthritis	Seizures
0 — 10	0 — 10	0 — 10	0 — 10

Nausea	Fatigue	Eye Pressure	Cramps
0 — 10	0 — 10	0 — 10	0 — 10

Other:	Other:	Other:	Other:
0 — 10	0 — 10	0 — 10	0 — 10

Taste

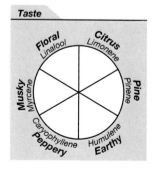

Description _____
_____
_____

Overall ___ /10

Recommend Yes / No

Experience _____
_____

Date _____

Dispensary (name/location) _____

Strain _____

indica ☐
sativa ☐

THC _____ **CBD** _____

Price _____ $/g

Scent

- Floral — Linalool
- Citrus — Limonene
- Pine — Pinene
- Earthy — Humulene
- Peppery — Caryophyllene
- Musky — Myrcene

Consumption Method

Flower   Concentrate   Edible   Topical

Description _____
_____
_____

Pain	Headache	Anxiety/Stress	Muscle Spasms
0 — 10	0 — 10	0 — 10	0 — 10

Appetite	Depression	Arthritis	Seizures
0 — 10	0 — 10	0 — 10	0 — 10

Nausea	Fatigue	Eye Pressure	Cramps
0 — 10	0 — 10	0 — 10	0 — 10

Other:	Other:	Other:	Other:
0 — 10	0 — 10	0 — 10	0 — 10

Taste

- Floral — Linalool
- Citrus — Limonene
- Pine — Pinene
- Earthy — Humulene
- Peppery — Caryophyllene
- Musky — Myrcene

Description _____
_____
_____

Overall _____ /10

Recommend Yes / No

Experience _____
_____
_____

Date _____

Dispensary (name/location) _____

Strain _____ indica ☐ sativa ☐

THC	CBD

Price _____ $/g

Scent

- Floral — Linalool
- Citrus — Limonene
- Pine — Pinene
- Earthy — Humulene
- Peppery — Caryophyllene
- Musky — Myrcene

Consumption Method

Flower Concentrate Edible Topical

Description _____

Pain	Headache	Anxiety/Stress	Muscle Spasms
0 — 10	0 — 10	0 — 10	0 — 10

Appetite	Depression	Arthritis	Seizures
0 — 10	0 — 10	0 — 10	0 — 10

Nausea	Fatigue	Eye Pressure	Cramps
0 — 10	0 — 10	0 — 10	0 — 10

Other:	Other:	Other:	Other:
0 — 10	0 — 10	0 — 10	0 — 10

Taste

- Floral — Linalool
- Citrus — Limonene
- Pine — Pinene
- Earthy — Humulene
- Peppery — Caryophyllene
- Musky — Myrcene

Description _____

Overall ____ /10

Recommend Yes / No

Experience _____

Date _____

Dispensary (name/location) _____

Strain _____ indica ☐ sativa ☐

THC _____ **CBD** _____

Price _____ $/g

Scent

Floral — Linalool
Citrus — Limonene
Musky — Myrcene
Pine — Pinene
Caryophyllene — Peppery
Humulene — Earthy

Consumption Method

Flower   Concentrate   Edible   Topical

Description _____
_____
_____

Pain	Headache	Anxiety/Stress	Muscle Spasms
0 — 10	0 — 10	0 — 10	0 — 10

Appetite	Depression	Arthritis	Seizures
0 — 10	0 — 10	0 — 10	0 — 10

Nausea	Fatigue	Eye Pressure	Cramps
0 — 10	0 — 10	0 — 10	0 — 10

Other:	Other:	Other:	Other:
0 — 10	0 — 10	0 — 10	0 — 10

Taste

Floral — Linalool
Citrus — Limonene
Musky — Myrcene
Pine — Pinene
Caryophyllene — Peppery
Humulene — Earthy

Description _____
_____
_____

Overall _____ /10

Recommend Yes / No

Experience _____
_____

Date

Dispensary (name/location)

Strain indica ☐ sativa ☐

THC   CBD

Price $/g

Scent

Floral — Linalool
Citrus — Limonene
Pine — Pinene
Earthy — Humulene
Peppery — Caryophyllene
Musky — Myrcene

Consumption Method

Flower **Concentrate** **Edible** **Topical**

Description

Pain	Headache	Anxiety/Stress	Muscle Spasms
0 — 10	0 — 10	0 — 10	0 — 10
Appetite	**Depression**	**Arthritis**	**Seizures**
0 — 10	0 — 10	0 — 10	0 — 10
Nausea	**Fatigue**	**Eye Pressure**	**Cramps**
0 — 10	0 — 10	0 — 10	0 — 10
Other:	Other:	Other:	Other:
0 — 10	0 — 10	0 — 10	0 — 10

Taste

Floral — Linalool
Citrus — Limonene
Pine — Pinene
Earthy — Humulene
Peppery — Caryophyllene
Musky — Myrcene

Description

Overall   /10

Recommend   Yes / No

Experience

Date _____

Dispensary (name/location) _____

Strain _____ indica ☐   **THC** _____ **CBD** _____
                                     sativa ☐   **Price** _____ $/g

Scent

- Floral — Linalool
- Citrus — Limonene
- Pine — Pinene
- Humulene — Earthy
- Caryophyllene — Peppery
- Myrcene — Musky

Consumption Method

Flower   Concentrate   Edible   Topical

Description _____
_____
_____

Pain	Headache	Anxiety/Stress	Muscle Spasms
0 — 10	0 — 10	0 — 10	0 — 10

Appetite	Depression	Arthritis	Seizures
0 — 10	0 — 10	0 — 10	0 — 10

Nausea	Fatigue	Eye Pressure	Cramps
0 — 10	0 — 10	0 — 10	0 — 10

Other:	Other:	Other:	Other:
0 — 10	0 — 10	0 — 10	0 — 10

Taste

- Floral — Linalool
- Citrus — Limonene
- Pine — Pinene
- Humulene — Earthy
- Caryophyllene — Peppery
- Myrcene — Musky

Description _____
_____

Overall _____ /10

Recommend Yes / No

Experience _____
_____

Date _____

Dispensary (name/location) _____

Strain _____ indica ☐ sativa ☐

THC _____ **CBD** _____

Price _____ $/g

Scent

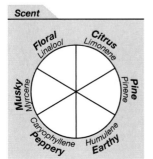

Consumption Method

Flower **Concentrate** **Edible** **Topical**

Description _____
_____
_____

Pain	Headache	Anxiety/Stress	Muscle Spasms
0 — 10	0 — 10	0 — 10	0 — 10

Appetite	Depression	Arthritis	Seizures
0 — 10	0 — 10	0 — 10	0 — 10

Nausea	Fatigue	Eye Pressure	Cramps
0 — 10	0 — 10	0 — 10	0 — 10

Other:	Other:	Other:	Other:
0 — 10	0 — 10	0 — 10	0 — 10

Taste

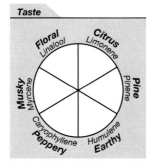

Description _____
_____
_____

Overall _____ /10

Recommend Yes / No

Experience _____
_____

Date _____

Dispensary (name/location) _____

Strain _____ indica ☐ sativa ☐

THC _____ **CBD** _____

Price _____ $/g

Scent

- Floral — Linalool
- Citrus — Limonene
- Pine — Pinene
- Earthy — Humulene
- Peppery — Caryophyllene
- Musky — Myrcene

Consumption Method

Flower **Concentrate** **Edible** **Topical**

Description _____
_____
_____

Pain	Headache	Anxiety/Stress	Muscle Spasms
0 — 10	0 — 10	0 — 10	0 — 10

Appetite	Depression	Arthritis	Seizures
0 — 10	0 — 10	0 — 10	0 — 10

Nausea	Fatigue	Eye Pressure	Cramps
0 — 10	0 — 10	0 — 10	0 — 10

Other:	Other:	Other:	Other:
0 — 10	0 — 10	0 — 10	0 — 10

Taste

- Floral — Linalool
- Citrus — Limonene
- Pine — Pinene
- Earthy — Humulene
- Peppery — Caryophyllene
- Musky — Myrcene

Description _____
_____
_____

Overall _____ /10

Recommend Yes / No

Experience _____
_____

Date _____

Dispensary (name/location) _____

Strain _____ indica ☐ sativa ☐

THC ____ **CBD** ____

Price ____ $/g

Scent

Floral — Linalool
Citrus — Limonene
Pine — Pinene
Earthy — Humulene
Peppery — Caryophyllene
Musky — Myrcene

Consumption Method

Flower   Concentrate   Edible   Topical

Description _____
_____
_____

Pain	Headache	Anxiety/Stress	Muscle Spasms
0 — 10	0 — 10	0 — 10	0 — 10

Appetite	Depression	Arthritis	Seizures
0 — 10	0 — 10	0 — 10	0 — 10

Nausea	Fatigue	Eye Pressure	Cramps
0 — 10	0 — 10	0 — 10	0 — 10

Other:	Other:	Other:	Other:
0 — 10	0 — 10	0 — 10	0 — 10

Taste

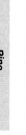

Floral — Linalool
Citrus — Limonene
Pine — Pinene
Earthy — Humulene
Peppery — Caryophyllene
Musky — Myrcene

Description _____
_____
_____

Overall ____ /10

Recommend Yes / No

Experience _____
_____

Date _____

Dispensary (name/location) _____

| THC _____ | CBD _____ |

Strain _____ indica ☐ sativa ☐   *Price* _____ $/g

Scent

- Floral — Linalool
- Citrus — Limonene
- Pine — Pinene
- Earthy — Humulene
- Peppery — Caryophyllene
- Musky — Myrcene

Consumption Method

Flower **Concentrate** **Edible** **Topical**

Description _____
_____
_____

Pain	Headache	Anxiety/Stress	Muscle Spasms
0 — 10	0 — 10	0 — 10	0 — 10

Appetite	Depression	Arthritis	Seizures
0 — 10	0 — 10	0 — 10	0 — 10

Nausea	Fatigue	Eye Pressure	Cramps
0 — 10	0 — 10	0 — 10	0 — 10

Other:	Other:	Other:	Other:
0 — 10	0 — 10	0 — 10	0 — 10

Taste

- Floral — Linalool
- Citrus — Limonene
- Pine — Pinene
- Earthy — Humulene
- Peppery — Caryophyllene
- Musky — Myrcene

Description _____
_____

Overall _____ /10

Recommend Yes / No

Experience _____
_____

Date _____

Dispensary (name/location) _____

Strain _____ indica ☐ sativa ☐

THC	CBD

Price _____ $/g

Scent

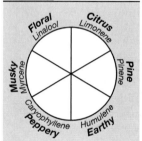

Consumption Method

Flower **Concentrate** **Edible** **Topical**

Description _____
_____
_____

Pain	Headache	Anxiety/Stress	Muscle Spasms
0 10	0 10	0 10	0 10

Appetite	Depression	Arthritis	Seizures
0 10	0 10	0 10	0 10

Nausea	Fatigue	Eye Pressure	Cramps
0 10	0 10	0 10	0 10

Other:	Other:	Other:	Other:
0 10	0 10	0 10	0 10

Taste

Description _____
_____
_____

Overall _____ /10

Recommend Yes / No

Experience _____
_____

Date _____

Dispensary (name/location) _____

Strain _____ indica ☐ sativa ☐

THC ____ CBD ____

Price ____ $/g

Scent

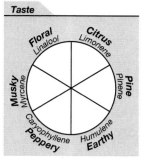

Consumption Method

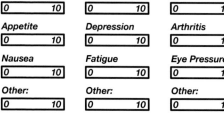

Flower   Concentrate   Edible   Topical

Description _____
_____
_____

Pain	Headache	Anxiety/Stress	Muscle Spasms
0 10	0 10	0 10	0 10

Appetite	Depression	Arthritis	Seizures
0 10	0 10	0 10	0 10

Nausea	Fatigue	Eye Pressure	Cramps
0 10	0 10	0 10	0 10

Other:	Other:	Other:	Other:
0 10	0 10	0 10	0 10

Taste

Description _____
_____
_____

Overall ____ /10

Recommend   Yes / No

Experience _____
_____

Date _____

Dispensary (name/location) _____

Strain _____ indica ☐ sativa ☐

THC _____ *CBD* _____

Price _____ $/g

Scent

Floral — Linalool
Citrus — Limonene
Pine — Pinene
Earthy — Humulene
Peppery — Caryophyllene
Musky — Myrcene

Consumption Method

Flower Concentrate Edible Topical

Description _____
_____
_____

Pain	Headache	Anxiety/Stress	Muscle Spasms
0 — 10	0 — 10	0 — 10	0 — 10

Appetite	Depression	Arthritis	Seizures
0 — 10	0 — 10	0 — 10	0 — 10

Nausea	Fatigue	Eye Pressure	Cramps
0 — 10	0 — 10	0 — 10	0 — 10

Other:	Other:	Other:	Other:
0 — 10	0 — 10	0 — 10	0 — 10

Taste

Floral — Linalool
Citrus — Limonene
Pine — Pinene
Earthy — Humulene
Peppery — Caryophyllene
Musky — Myrcene

Description _____
_____
_____

Overall _____ /10

Recommend Yes / No

Experience _____
_____

Date _____

Dispensary (name/location) _____

| THC | CBD |

Strain _____ indica ☐ sativa ☐   Price _____ $/g

Scent

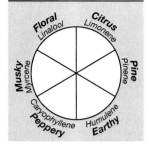

Consumption Method

Flower **Concentrate** **Edible** **Topical**

Description _____
_____
_____

Pain	Headache	Anxiety/Stress	Muscle Spasms
0 — 10	0 — 10	0 — 10	0 — 10
Appetite	**Depression**	**Arthritis**	**Seizures**
0 — 10	0 — 10	0 — 10	0 — 10
Nausea	**Fatigue**	**Eye Pressure**	**Cramps**
0 — 10	0 — 10	0 — 10	0 — 10
Other:	**Other:**	**Other:**	**Other:**
0 — 10	0 — 10	0 — 10	0 — 10

Taste

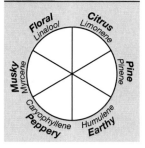

Description _____
_____
_____

Overall _____ /10

Recommend   Yes / No

Experience _____
_____

Date _____

Dispensary (name/location) _____

Strain _____ indica ☐ sativa ☐

THC	CBD

Price _____ $/g

Scent

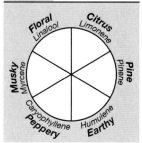

Consumption Method

Flower   Concentrate   Edible   Topical

Description _____
_____
_____

Pain	Headache	Anxiety/Stress	Muscle Spasms
0 10	0 10	0 10	0 10

Appetite	Depression	Arthritis	Seizures
0 10	0 10	0 10	0 10

Nausea	Fatigue	Eye Pressure	Cramps
0 10	0 10	0 10	0 10

Other:	Other:	Other:	Other:
0 10	0 10	0 10	0 10

Taste

Description _____
_____
_____

Overall _____ /10

Recommend   Yes / No

Experience _____
_____

Date _____

Dispensary (name/location) _____

Strain _____ indica ☐ sativa ☐

THC _____ CBD _____

Price _____ $/g

Scent

- Floral — Linalool
- Citrus — Limonene
- Pine — Pinene
- Earthy — Humulene
- Peppery — Caryophyllene
- Musky — Myrcene

Consumption Method

Flower Concentrate Edible Topical

Description _____

Pain	Headache	Anxiety/Stress	Muscle Spasms
0 — 10	0 — 10	0 — 10	0 — 10

Appetite	Depression	Arthritis	Seizures
0 — 10	0 — 10	0 — 10	0 — 10

Nausea	Fatigue	Eye Pressure	Cramps
0 — 10	0 — 10	0 — 10	0 — 10

Other:	Other:	Other:	Other:
0 — 10	0 — 10	0 — 10	0 — 10

Taste

- Floral — Linalool
- Citrus — Limonene
- Pine — Pinene
- Earthy — Humulene
- Peppery — Caryophyllene
- Musky — Myrcene

Description _____

Overall _____ /10

Recommend   Yes / No

Experience _____

Date _____

Dispensary (name/location) _____

Strain _____
indica ☐
sativa ☐

THC _____ **CBD** _____
Price _____ $/g

Scent

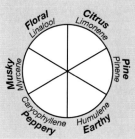

Consumption Method

Flower Concentrate Edible Topical

Description _____

Pain	Headache	Anxiety/Stress	Muscle Spasms
0 — 10	0 — 10	0 — 10	0 — 10
Appetite	**Depression**	**Arthritis**	**Seizures**
0 — 10	0 — 10	0 — 10	0 — 10
Nausea	**Fatigue**	**Eye Pressure**	**Cramps**
0 — 10	0 — 10	0 — 10	0 — 10
Other:	**Other:**	**Other:**	**Other:**
0 — 10	0 — 10	0 — 10	0 — 10

Taste

Description _____

Overall _____ /10

Recommend Yes / No

Experience _____

Date _____

Dispensary (name/location) _____

Strain _____

indica ☐
sativa ☐

THC _____ **CBD** _____

Price _____ $/g

Scent

- Floral — Linalool
- Citrus — Limonene
- Pine — Pinene
- Earthy — Humulene
- Peppery — Caryophyllene
- Musky — Myrcene

Consumption Method

Flower · Concentrate · Edible · Topical

Description _____
_____
_____

Pain	Headache	Anxiety/Stress	Muscle Spasms
0 — 10	0 — 10	0 — 10	0 — 10

Appetite	Depression	Arthritis	Seizures
0 — 10	0 — 10	0 — 10	0 — 10

Nausea	Fatigue	Eye Pressure	Cramps
0 — 10	0 — 10	0 — 10	0 — 10

Other:	Other:	Other:	Other:
0 — 10	0 — 10	0 — 10	0 — 10

Taste

- Floral — Linalool
- Citrus — Limonene
- Pine — Pinene
- Earthy — Humulene
- Peppery — Caryophyllene
- Musky — Myrcene

Description _____
_____
_____

Overall _____ /10

Recommend Yes / No

Experience _____
_____

Date

Dispensary (name/location)

	THC	CBD

Strain — indica ☐ sativa ☐ *Price* $/g

Scent

- Floral / Linalool
- Citrus / Limonene
- Musky / Myrcene
- Pine / Pinene
- Caryophyllene / Peppery
- Humulene / Earthy

Consumption Method

Flower **Concentrate** **Edible** **Topical**

Description

Pain	Headache	Anxiety/Stress	Muscle Spasms
0 — 10	0 — 10	0 — 10	0 — 10

Appetite	Depression	Arthritis	Seizures
0 — 10	0 — 10	0 — 10	0 — 10

Nausea	Fatigue	Eye Pressure	Cramps
0 — 10	0 — 10	0 — 10	0 — 10

Other:	Other:	Other:	Other:
0 — 10	0 — 10	0 — 10	0 — 10

Taste

- Floral / Linalool
- Citrus / Limonene
- Musky / Myrcene
- Pine / Pinene
- Caryophyllene / Peppery
- Humulene / Earthy

Description

Overall /10

Recommend Yes / No

Experience

Date _____

Dispensary (name/location) _____

Strain _____ indica ☐ sativa ☐

THC ____ **CBD** ____ **Price** ____ $/g

Scent

- Floral — Linalool
- Citrus — Limonene
- Pine — Pinene
- Earthy — Humulene
- Peppery — Caryophyllene
- Musky — Myrcene

Consumption Method

Flower Concentrate Edible Topical

Description _____
_____
_____

Pain	Headache	Anxiety/Stress	Muscle Spasms
0 — 10	0 — 10	0 — 10	0 — 10

Appetite	Depression	Arthritis	Seizures
0 — 10	0 — 10	0 — 10	0 — 10

Nausea	Fatigue	Eye Pressure	Cramps
0 — 10	0 — 10	0 — 10	0 — 10

Other:	Other:	Other:	Other:
0 — 10	0 — 10	0 — 10	0 — 10

Taste

- Floral — Linalool
- Citrus — Limonene
- Pine — Pinene
- Earthy — Humulene
- Peppery — Caryophyllene
- Musky — Myrcene

Description _____
_____
_____

Overall ____ /10

Recommend Yes / No

Experience _____
_____

ADDITIONAL NOTES:

Made in United States
North Haven, CT
22 June 2023